Water
and Climate

Full Option Science System
Developed at
The Lawrence Hall of Science,
University of California, Berkeley
Published and distributed by
Delta Education,
a member of the School Specialty Family

1487705
978-1-62571-324-7
Printing 3 — 7/2015
Quad/Graphics, Versailles, KY

Table of Contents

A Report from the Blue Planet

TO: Chief of Science, Home Planet
FROM: Interplanetary Science Office, Fleet 2087

Greetings from the blue planet mentioned in my last report. We have been exploring the planet as directed. Now we know why it looks blue from space. Almost three-quarters of the planet's surface is covered by **water**! In all our planet explorations, this is the first water planet we have discovered. Here's what we have learned so far.

Ninety-seven percent of the planet's water is in its huge ocean of **salt water**. Our first view of the blue planet was almost all ocean. When we flew around to the other side, we saw that there is dry land, too.

A view of the Atlantic Ocean

A view of the Pacific Ocean

The rest of the planet's water is **fresh water**. That means only 3 percent of the water is free of salt. And about two-thirds of the fresh water is **solid ice**. That leaves just 1 percent of the planet's water as **liquid** fresh water.

The ocean makes up 97 percent of Earth's water.

Liquid fresh water is found in many places. A lot of the water is underground. The rest of the fresh water is on the planet's solid surface. We see it in lakes and rivers. All the plants and animals on the blue planet need water to stay alive. The people living there use water in many ways. They use it for cooking, washing, drinking, and growing food.

We have observed water in three states on the planet. It is the only material found naturally on this planet in all three states. Water can be solid ice, liquid water, and a **gas** called **water vapor**.

Water vapor is in the air. There is more water in the air than in all the rivers on the planet. We will find out more about water vapor for our next report.

As you can see, water is an amazing material. It is in the ocean, in lakes and rivers, in the ground, and in the air. It is everywhere.

Earth's Water

97%

2%   1%

Ice   Liquid

Salt Water   Fresh Water

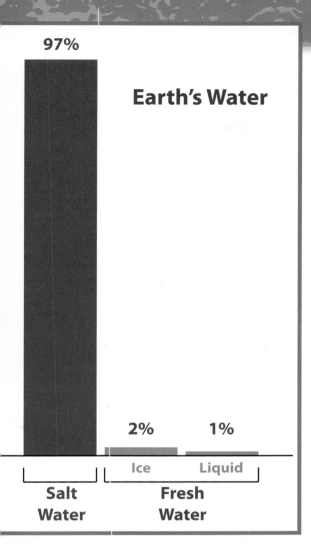

Two percent of Earth's fresh water is found in solid ice.

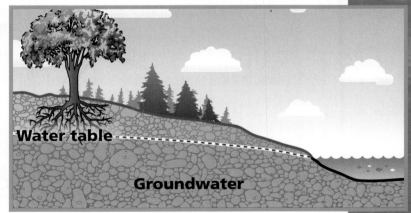

Water table

Groundwater

The remaining 1 percent is liquid fresh water found in Earth's lakes, rivers, and groundwater.

Surface Tension

Have you ever seen an insect walk on water? If you have, you may have wondered, how can it do that? The answer is **surface tension**. Water is made of tiny particles. The particles are naturally attracted to one another. At the surface, where water meets the air, the attraction between particles is very strong. The strong attraction at the surface of the water is surface tension. Insects like water striders can walk on water because bristles on their feet keep them from breaking through the surface tension.

You can see how surface tension works. Fill a glass to the top with water. Keep adding more water a little at a time. If you are careful, you can "overfill" the glass. The water will form a dome above the top of the glass, but it won't spill out. Why? Surface tension.

What happens when water falls through air? Water forms drops. Drops are small **volumes** of water with surface all around. The skinlike surface tension pulls all around the outside of the drop. The pulling results in a sphere. Next time you are in the shower, look closely at the falling water. What do you see?

Without surface tension, **rain** falling from **clouds** might fall in sheets or strings. Without surface tension, water landing on a window, a car, or a leaf would spread out into a thin film. But the water doesn't make a thin film. It forms dome-shaped bits of water called beads. Why do you think water forms beads when it falls on a waterproof surface?

Remember surface tension the next time you watch water striders zip across the water. The little insects glide over water as if they were skating on ice because of surface tension.

Water forms beads.

Which Way Does It Go?

Go outside during a rainstorm and look around. What happens to the rainwater? Some of it **soaks** down into the **soil**. Some of it flows across the ground, sidewalks, and paved surfaces. Water always seems to be on the move. Why is that? Let's follow a few drops of water that are on the move.

Look on top of the mountain. There is still some snow high up around the peaks. Drops of **melted** snow flow down the sides of mountains and into brooks. Brooks join to form streams, and streams tumble over cliffs as waterfalls.

Streams flow into rivers. Drops of water in rivers slow down when the river is dammed. But they don't stop. When water drops pass over the dam, they flow to the ocean. When water gets to the ocean, it finally stops moving. Or does it? There may be more to the story of a water drop.

Look again at the pictures. Follow the water drops from the mountain peak to the ocean. Which way does water go? Water always flows in the same direction. Water always moves down.

Water is **matter**. Like all matter, water is pulled downward by **gravity**. That's why brooks flow from mountain peaks to forest meadows. That's why meadow streams flow into river valleys. That's why rivers flow down to the ocean.

The next time it rains, watch the water flowing across the ground. Which way is it going?

Opinion and Evidence

Two girls just finished a sponge activity. They were surprised that their 4-**gram (g)** sponge soaked up 32 g of water. That seems like a lot for such a small sponge.

As they recorded data in notebooks, Teasha said, "If we had a natural sponge, it would soak up even more water."

"How do you know?" asked Kim.

"I just know it would," replied Teasha. "Natural things are always better. I would always choose a natural sponge. I'm sure it would work better."

"So you've never tested a natural sponge to find out if it can soak up more water than a synthetic sponge?" asked Kim.

"Well, no, I never actually did the experiment," admitted Teasha. "But it makes sense to me that the natural sponge would soak up water better."

"We could find out for sure," said Kim. "Let's get a natural sponge and test it. That should give us **evidence** about your **opinion** that natural things are better."

A natural sponge

A synthetic sponge

The Experiment

The next day, the girls stayed after school to do their experiment. They had a new synthetic sponge and a new natural sponge. But there was a problem. The natural sponge was much larger than the synthetic one.

Teasha and Kim decided to cut three small samples from each sponge. The small samples would all be the same shape and same **mass**. They cut and trimmed and **weighed** carefully. Finally, all six samples were exactly 5 g.

"How should we soak the sponges to make sure it is a fair experiment?" asked Kim.

"I know," said Teasha. "We can use a stopwatch to time 1 minute while we hold the sponges under water. That will really soak the sponges. Then we'll take them out of the water. We will hold them over the basin for 30 seconds. Then we will weigh them to find out how much water soaked into each sponge."

"That sounds good to me," agreed Kim. "Let's get started."

The girls soaked and weighed the first synthetic sponge. They repeated the procedure with the other two synthetic sponges. They did this to make sure their measurements were accurate. Then they did the same thing with the three natural sponges. They recorded their measurements in a table.

The girls soaked the sponges for 1 minute.

Then they let the sponges drip for 30 seconds.

Sponge	Mass of sponge (g)	Mass of wet sponge (g)	Mass of water (g)
Synthetic 1	5	45	40
Synthetic 2	5	46	41
Synthetic 3	5	45	40
Natural 1	5	41	36
Natural 2	5	40	35
Natural 3	5	39	34

A Second Look

The girls studied the data. It looked like the synthetic sponge soaked up about 5 more grams of water than the natural sponge.

"Hmmm," said Teasha, "it looks like the natural sponge isn't better, at least not better at soaking up water. But you know what? I want to try one more thing. Let's squeeze as much water out of the sponges as we can. Then, starting with the damp sponges, we will repeat the experiment exactly. Then we will be sure our results are accurate."

Kim thought that was a good idea. They repeated the experiment and recorded these data.

Sponge	Mass of sponge (g)	Mass of wet sponge (g)	Mass of water (g)
Synthetic 1	7	45	38
Synthetic 2	8	46	38
Synthetic 3	8	45	37
Natural 1	7	41	34
Natural 2	8	40	32
Natural 3	7	39	32

"OK, I see now that the synthetic sponge is better at soaking up water," said Teasha. "The evidence is right there for all to see. From now on, I am going to use synthetic sponges to soak up spills. But I will still use natural sponges for other things because they last longer."

"Are you sure?" asked Kim.

Opinion

Teasha likes natural things. She likes chairs made of wood. She likes T-shirts made of cotton. Her opinion is that natural things are always better.

When she and Kim were working with sponges, Teasha claimed that natural sponges were better. But her claim was not based on data and evidence. Her claim was her opinion. Opinions are based on what a person believes to be true, not on scientific data. Evidence is based on observation and scientific data.

In science, claims are tested with experiments. Experiments produce data and evidence. The evidence will show if the claim is true or not true. Sometimes more experiments need to be done before a conclusion can be reached. When Teasha and Kim did their experiment, the evidence showed that the synthetic sponge soaked up more water. Teasha changed her mind about sponges after she studied the evidence.

Thinking about Opinions and Evidence

1. Teasha claimed that natural sponges were better. What did she base that claim on?

2. Why did Teasha and Kim repeat their experiment?

3. Was Teasha's claim that natural sponges last longer based on opinion or evidence?

4. What is the difference between opinion and evidence?

Water Everywhere

It's easy to take water for granted. Water is everywhere. It's the most common substance on our planet. Water covers more than 70 percent of Earth. But only about 1 percent of all Earth's water is fresh water that people can use.

Water is one of Earth's most precious **natural resources**. All living things need water to survive. In some areas of the world, water is scarce and more valuable than gold. Even in parts of the United States, **droughts** and water shortages can occur.

How Much Water Do We Use?

- Each American uses about 300 to 380 **liters (L)** of water each day.
- Flushing the toilet uses between 15 and 26 L each time, depending on the type of toilet.
- A bath uses about 114 L.
- A shower uses between 19 and 38 L per minute.
- Do you leave the water running when you brush your teeth? If so, you use from 3 to 7 L of water each time.
- A dripping faucet can waste more than 3,800 L of water each year.

Be a Water Watcher

What can you do to **conserve** water? An easy way to conserve water is to pay attention to how much water you use each day. Here are some other tips to help you become a water watcher.

- Keep a pitcher of water in the refrigerator. Then you won't have to run the faucet to get really cold water.

- Take short showers instead of baths.

- Use low-flow faucets and showerheads.

- Don't let the water run while you brush your teeth. Also, turn off running water while you soap up your hands.

- Don't throw facial tissues and other trash into the toilet. Use a trash can instead. This will stop clogs and cut down on the number of times you flush.

- If you have a fish tank, **recycle** the water by giving it to your plants. The fish-tank water is a good plant fertilizer.

A short shower uses less water than a bath.

Fill 'er Up

Get the lowdown on these amazing measurements!
- Elephants need a lot of water. They can drink from 75 to 100 L each day.
- The average American drinks about 19 L of orange juice each year. Nine out of every ten Florida oranges are squeezed into juice.
- The average American eats more than 24 L of ice cream a year!
- Camels are prized animals in the desert. They can go for long stretches without any water. A camel that has gone without water for a long time can drink 100 L or more at once.

Vacation Aggravation

January 3

Dear Grandma and Grandpa,

We just arrived in Sydney, Australia. Wow, did I get a big surprise! I had read that the January temperatures in some parts of Australia were usually around 28°. I packed all my warmest winter clothes after I read that. When I got off the plane, it was unbelievably hot! I thought it was some kind of weird heat wave!

I made a BIG mistake. Temperatures in the northern part of Australia do average 28° in January. But that's 28°C! That's about 83°F. It turns out that because Australia is in the southern half of the world, their summer is our winter. Mom's still pretty mad. She had to buy me a bunch of shorts and shirts to wear. She says that's the last time she'll ever let me pack my own suitcase!

Love, Ami

Here's the opera house in Sydney.

January 5

Hi Grandma and Grandpa,

Today Mom and I visited the Taronga Zoo in Sydney. I couldn't wait to see the koalas.

I had read they weigh 14 kilograms. That's about 30 pounds. Did you know that koalas are endangered? Today many live in zoos.

Love, Ami

January 8

Hi Grandma and Grandpa,

Today Mom and I visited Ayers Rock. Before I got here, my friend Bill had told me that the rock was 345 feet high. I've climbed that high before, so I was excited to climb Ayers Rock. When we got here, though, Ayers Rock turned out to be 345 meters high! That's over 1,140 feet. We arrived too late in the day to climb to the top, so Mom and I enjoyed the view from the bottom.

Love, Ami

January 10

Hi Grandma and Grandpa,

Today Mom and I took a ride through the Australian countryside. We saw some kangaroos. Here's a weird fact. Australians drive on the wrong side of the road. Yep, everyone down under drives on the left. That made me nervous. When I saw that the speed limit was 100, I got VERY nervous. Then I figured out that 100 kilometers per hour is only about 62 miles per hour. After that, I could sit back and relax.

Love, Ami

January 12

G'day!

Today we checked out one of western Australia's beautiful beaches. It was terrific! And this time, I was prepared. I knew that the water temperature was a warm 20°C (about 68°F). I also knew that the walk to the beach from the hotel was 3 kilometers (1.8 miles). Now that I know that Australia, like most other countries, uses the metric system to measure things, I don't feel so out of place. Mom likes it here, too. In Australia, she weighs only 70 kilograms. (Do you know how many pounds that is?) See you in a fortnight (that's 2 weeks).

Love, Ami

Metric-to-English Conversions

These common conversions might have helped Ami while she was traveling in Australia.

- A centimeter is about half an inch.
- A meter is a little more than 3 feet, or 1 yard.
- A kilometer is about 0.6 miles.
- A kilogram is a little more than 2 pounds.
- A liter is about 1 quart.

Celsius and Fahrenheit

Celsius and Fahrenheit are two **scales** used to **measure temperature**. Both scales are based on the **freezing point** and **boiling point** of pure water at sea level. The **Celsius** scale has 100° between the two points. The Fahrenheit scale has 180° between the freezing point and boiling point.

Today most countries use the Celsius scale to measure temperatures. The United States, however, still uses the Fahrenheit scale.

Celsius

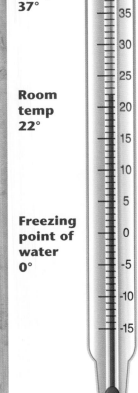

Body temp 37°

Room temp 22°

Freezing point of water 0°

Anders Celsius

The Celsius scale is named for Anders Celsius, a Swedish astronomer. Celsius lived from 1701 to 1744. In 1742, he created a temperature scale. This scale used 0°C to mark water's boiling point and 100°C to mark its freezing point. A few years later, another scientist changed Celsius's scale so that 0°C was the freezing temperature and 100°C was the boiling temperature. Celsius's scale was originally called the centigrade scale. It was renamed in the 1940s to honor the inventor.

Daniel G. Fahrenheit

The Fahrenheit scale is named for German scientist Daniel G. Fahrenheit. Fahrenheit lived from 1686 to 1736. In 1714, he invented the first mercury **thermometer**. He invented a temperature scale to go along with it. Fahrenheit's thermometer marked normal human body temperature as 98.6°F.

Fahrenheit thought he had found the lowest possible temperature by mixing ice and salt. He set the temperature of this **mixture** at 0°F. Then he set the freezing point of water at 32°F. He also set the boiling point of water at 212°F.

Fahrenheit

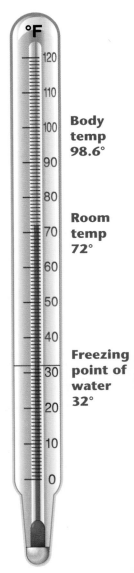

Body temp 98.6°

Room temp 72°

Freezing point of water 32°

Water: Hot and Cold

When things get hot, something interesting happens. They get bigger. Usually you can't see that the hot material is bigger. The change is small. But one place you can see that hot material is bigger is in a bulb thermometer.

A bulb thermometer is a small container of liquid attached to a thin tube. The small container is the bulb. The thin tube is the stem. When the bulb gets hot, the liquid **expands** (gets larger). Liquid pushes farther up the stem. When the bulb gets cold, the liquid **contracts** (gets smaller). Liquid pulls back into the bulb.

How does that happen? It happens at a level that is invisible to our eyes.

This is what scientists have figured out. Water is made of tiny particles that are much too small to see. The particles are moving around all the time. They move faster when the water is hot and slower when the water is cold.

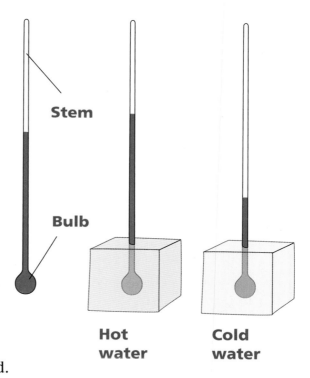

Stem

Bulb

Hot water   **Cold water**

Think about a pan of water. All the water particles bang into one another all the time. That keeps a little space between the particles. When the water is hot, the particles bang into one another harder. Harder banging pushes the particles a little farther apart. When the particles are farther apart, the volume of water in the pan increases. Increased volume is expansion.

Now can you explain what happens to the liquid in a bulb thermometer?

Particles of cold water in a pan

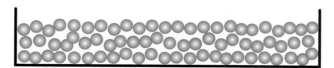

Particles of hot water in a pan

Float and Sink

Imagine that you are having sunflower seeds for a snack and that they spill. The seeds fall onto gravel where they are hard to see. How could you separate this mixture of seeds and gravel? Just scoop up the seeds and gravel and drop them into a bowl of water. The pieces of rock (gravel) will **sink**. The sunflower seeds will **float**.

Why do the seeds float and the bits of rock sink? Some might say it is because rocks are heavier than sunflower seeds. But that wouldn't be true.

Think about this. A piece of gravel on one side of a balance and a seed on the other side have the same mass. Each has a mass of exactly 0.1 gram (g).

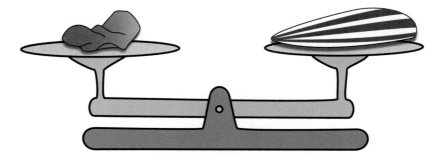

A small piece of gravel and a large sunflower seed have the same mass.

If we drop these two objects in water, the seed will still float, and the rock will still sink. Why? Because the volumes are different. The two objects have the same mass, but the mass is more concentrated in the piece of rock. The rock is **more dense** than the seed.

Density is the amount of mass compared to the volume. Imagine that we can scrunch both objects into perfect spheres. The mass will still be the same, but now we can compare the volumes.

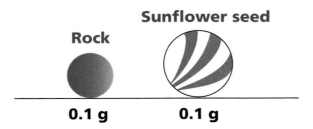

The rock has the same mass as the sunflower seed, but in a smaller volume, so the rock is more dense. But why does the rock sink and the seed float? Look at the same mass of water.

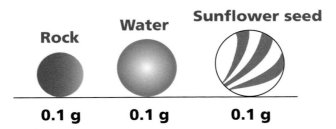

Compare the rock and sunflower-seed spheres to an equal mass of water. The volume of the water is larger than the volume of the rock, but smaller than the volume of the sunflower seed. The rock is more dense than the water, so it sinks. The sunflower seed is less dense than the water, so it floats.

Water Density

Why does warm water form a layer on top of room-temperature water? Water particles move faster when water gets hot. Particles push one another farther apart. The water expands. When water expands, the mass stays the same, but the volume increases. What happens to the density?

Look at the glass of layered water. Which is the hot water? Which is the cold water? How do you know?

Ice Is Everywhere

You probably know where to go to get some ice. Indoors you find a refrigerator and look in the freezer. Outdoors is a different story. If it's winter, and you live between Montana and Maine along the border with Canada, ice is everywhere. Every pond, creek, and bucket of water is frozen. In the warmer parts of the country, and during the summer, finding ice outdoors can be a challenge.

Some places are cold all year long. Alaska, Canada, Greenland, Iceland, Scandinavia, and Siberia have ice year-round. Antarctica, which covers the South Pole, is the iciest continent. More than 95 percent of its land lies under thick ice. In some places, the ice is 4,300 meters (m) thick. In the winter, frozen sea water forms an ice shelf around Antarctica doubling the continent's size!

If you live in snow country, you know what to expect. Usually starting in December, heavy snow falls, covering everything under a white blanket. During a heavy snow year, the snow may stay on the ground until March or April. Then it melts.

Polar bears on the ice in the Arctic

Ice off the coast of Greenland

Glaciers

What if the winter snow didn't melt during the summer? In some of the colder regions around the world, more snow falls than can melt in the summer. Snow piles up and up. The layers of snow at the bottom get compressed and turn into pure ice. When the ice is about 18 m thick, it begins to move. Moving ice is a **glacier**. Glaciers are "rivers" of ice that gravity pulls downhill.

Scientists can keep track of how fast glaciers move. An average glacier advances less than 1 m each day. A glacier in Greenland holds the **speed** record. Jakobshavn Glacier is speeding along at more than 35 m per day.

Glaciers now cover about 10 percent of Earth's land. They are found in all of the world's major mountain ranges. All the ice in the world store about 65 percent of the world's fresh water. If all that ice melted, sea level would rise about 79 m.

An Alaskan glacier

A glacier ends at the sea.

25

Icebergs

Icebergs are "mountains" of ice drifting like islands in the ocean. Icebergs are frozen fresh water, not salt water. Where does all the frozen fresh water come from?

Icebergs come from glaciers. When a glacier moving down a valley reaches the sea, pieces at the end break off. These chunks of ice may be as small as cars or as big as mountains. The largest iceberg ever measured was 320 kilometers (km) long!

We see only a small part of an iceberg. Seven-eighths is hidden beneath the water's surface. Icebergs in the North Atlantic can last up to 2 years before melting. Larger icebergs in the Antarctic may last 10 years.

Someday icebergs may be a useful source of fresh water. Ice could be harvested and melted. Whole icebergs might even be towed to countries needing fresh water!

Icebergs form when pieces of ice break away from the face of a glacier.

A large iceberg might extend a kilometer below the surface of the sea.

Ice History

Before refrigeration, people used ice to keep food cool. Ice was harvested from frozen lakes and rivers in winter. People waited until the ice was at least 60 centimeters (cm) thick. Then it was strong enough to hold the ice workers.

Horses were used to plow a frozen lake to clear away the snow. Then the horses pulled a special tool that scratched lines in the ice. Workers cut along the lines with sharp saws. They used poles to push large sheets of ice to icehouses. There they cut the sheets into smaller blocks.

Icehouses looked like barns. Inside, the ice workers carefully stacked and stored the ice blocks. They spread straw or sawdust over and around each block to keep it from melting and sticking to other ice blocks.

Ice workers mark lines on the frozen lake.

Large sheets of ice are pushed to the icehouse.

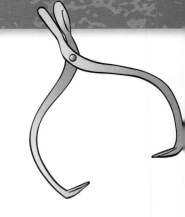

Most homes had an icebox. Throughout the year, horse-drawn trucks carried blocks of ice to homes in towns and cities. The iceman used ice tongs to handle the heavy blocks. Ice blocks could range from 11 to 22 kilograms (kg) each. The iceman then used an ice pick to fit the block of ice inside the icebox.

Children loved to see the ice truck on hot summer days. They crowded around when the iceman pushed the door open. There was sure to be a sliver of ice for each of them. What a treat on a steamy day!

Why Pipes Burst

Pipes supply water to houses, schools, and other buildings. They are made of strong materials, like plastic, copper, cast iron, and steel. But sometimes pipes burst! Do you know why?

Water, like all other materials, contracts as it cools. But once the temperature gets down to 4 degrees Celsius (°C), an amazing thing happens. As water continues to get colder, it starts to expand. Between 4°C and 0°C (the temperature at which water **freezes**), water expands. That means the ice needs more space as it changes from liquid to solid. If liquid water completely fills a container, it will break the container when it freezes. Ice needs room to expand, or it will break its container.

How can you prevent pipes from breaking in really cold **weather**? If the pipes are not used during the winter, drain the water out. If that is not possible, make sure pipes are well insulated. Pipes can be wrapped with insulation, or they can be buried deep underground.

Another solution is to leave the tap open just a little bit. Then the expanding water can push out the end of the pipe. The water might freeze in the pipe, but the pipe will not break. If the tap is closed, the water is trapped. Then something will break when the water freezes.

Remember those icebergs in the northern and southern seas? Why do they float? There is a connection between floating icebergs and breaking water pipes. Do you know what it is?

Why do pipes break? Because water expands as it turns to ice. The amount (mass) of water doesn't change, only its volume. If the mass stays the same but the volume increases, the density of ice changes. If ice is less dense than liquid water, what will happen when you put ice in water? It floats.

Studying Weather

Meteorologists are scientists who study the weather. Weather is the condition of the air in an area, and it is always changing. That is why **meteorologists** must constantly observe and measure those conditions. They use weather instruments to gather information so they can **predict** the weather. Meteorologists measure the temperature of the air. They observe cloud patterns. They measure how much rain or snow falls. They measure the speed and direction of the wind.

Temperature

Temperature is a measure of how hot the air is. Temperature is measured with a thermometer. There are many kinds of thermometers. The most common kind is a liquid thermometer. A liquid thermometer is a thin glass tube connected to a small bulb of liquid. As the liquid warms and cools, it expands and contracts. The height of the column of liquid in the tube changes in response to the temperature. By labeling the liquid tube to show temperatures, the meteorologist can read the temperature directly from the thermometer.

Metals also expand and contract in response to temperature change. Some thermometers use strips made of two different metals to detect temperature changes. These are called bimetallic thermometers. The two metals have different rates of expansion. One side of the strip expands more than the other as it **heats** up, and the strip bends. A pointer on the end of the bending strip points to the temperature.

A liquid thermometer

Precipitation

Some clouds bring rain or snow. Water in any form that falls to Earth from clouds is called **precipitation**. Precipitation is measured using a rain gauge. The kind of precipitation that falls depends on how cold the air is.

Precipitation falls as rain when the air between the clouds and Earth's surface is warmer than 0 degrees Celsius (°C). Most precipitation falls as rain.

Sleet forms when rain passes through air that is cooler than 0°C. Because the temperature is below 0°C, the raindrops freeze, forming bits of ice.

Hail forms in thunderstorm clouds especially when there are strong winds blowing large droplets of water upward. To form hail, a large part of the cloud has to be below 0°C.

Snow falls from clouds made of tiny ice crystals. If the air between the clouds and Earth is cooler than 0°C, the ice crystals do not melt as they fall.

A rain gauge

Wind Speed

Moving air is called wind. Meteorologists are interested in how fast the wind is moving. To measure wind speed, meteorologists use **anemometers** and **wind meters**. An anemometer uses a rotating **shaft** with wind-catching cups attached at the top. The harder the wind blows, the faster the shaft rotates, and the faster the cups move through the air. The moving cups measure the wind speed.

A wind meter is an instrument with a small ball in a tube. When wind blows across the top of the tube, the flow of air up the tube lifts the ball. The harder the wind blows, the higher the ball rises. Both instruments are adjusted to report wind in miles per hour (mph) or kilometers (km) per hour.

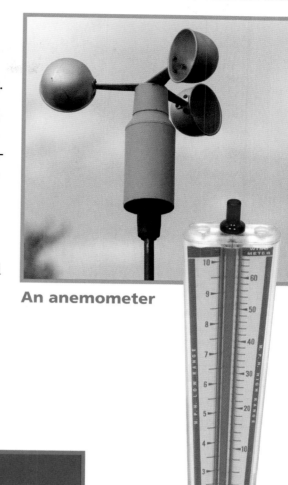

An anemometer

A wind meter

Plants and flags can show the direction the wind is blowing.

Wind Direction

Meteorologists are also interested in the direction the wind is blowing. To find out wind direction, meteorologists use a **wind vane**. A wind vane is a shaft with an arrow point on one end and a broad paddle shape at the other end. When wind hits the paddle, it rotates the shaft so that the arrow points into the wind. Using a **compass**, the meteorologist finds out the direction the shaft is pointing. Wind direction is the direction from which the wind is blowing. It is reported in compass directions, such as north or south.

A wind vane

A compass

Modern Weather Instruments

Meteorologists now use a combination of traditional weather instruments and computer-based digital weather instruments. Meteorologists get information from advanced electronic instruments that are placed in good locations for monitoring weather. Those instruments use radio transmitters (like those in cell phones) to send information to weather centers where meteorologists work.

This weather device for home use has electronic instruments inside for detecting and reporting temperature and **humidity**. Some models measure **air pressure** and are connected to anemometers to measure wind speed.

A digital weather instrument for home use

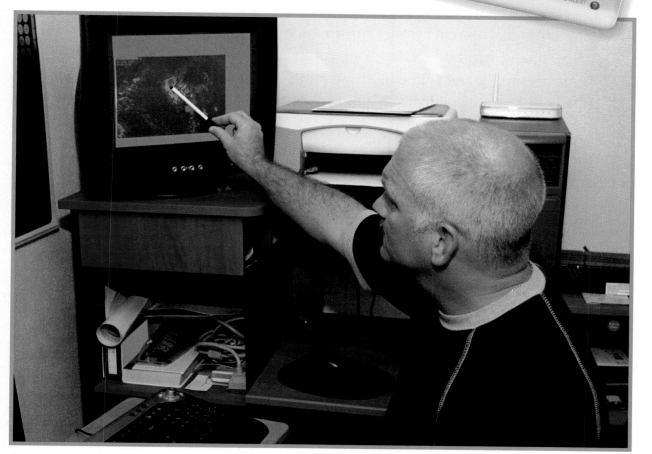

A meteorologist studies a radar image of a storm

Weather balloons carry weather instruments high into the sky. The weather instruments gather information about air temperature, wind speed, and wind direction. They gather information about air pressure. Air pressure is the **force** of air pushing on things around it. Weather balloons also provide information about humidity. Humidity is the amount of water in the air.

All this information helps meteorologists predict what weather is coming. They can make **forecasts** that help people know what to expect. We need to know the weather to make choices about what to wear, how to travel, and what events to plan.

Thinking about Weather

1. Who are meteorologists and what do they do?

2. How do we measure air temperature? wind direction? precipitation?

3. What do meteorologists use weather balloons for?

4. Why is it important for meteorologists to be able to forecast the weather?

Drying Up

You know when something is wet. It is covered with water, or it has soaked up a lot of water. When it rains, everything outside gets wet. When you go swimming, you and your swimsuit get wet. Clothes are wet when they come out of the washer. A dog is wet after a bath.

But things don't stay wet forever. Things get dry, often by themselves. An hour or two after the rain stops, porches, sidewalks, and plants are dry. After a break from swimming to eat lunch, you and your swimsuit are dry. After a few hours on the clothesline, clothes are dry. A dog is dry and fluffy after a short time. Where does the water go?

You can't see water vapor in the sky.

The water **evaporates**. When water evaporates, it changes
from water in its liquid form to water in its gas form. The gas form
of water is called water vapor. The water vapor leaves the wet object
and goes into the air. As the water evaporates, the wet object gets dry.

What happens when you put a wet object in a sealed container?
It stays wet. If you put your wet swimsuit in a plastic bag, it's still wet
when you take it out of the bag. Why? A little bit of the water in your
suit evaporates, but it can't escape into the air. The water vapor has no
place to go, so your suit is still wet when you get home.

Have you ever seen water vapor in the air? No, water vapor is
invisible. When water changes into vapor, it changes into individual
water particles. Water particles are too small to see with your eyes.
The water particles move into the air among the nitrogen and oxygen
particles. When water becomes part of the air, it is no longer liquid
water. It is a gas called water vapor.

Surface-Area Experiment

Julie and Art want to find out how **surface area** affects evaporation. They decide to do an experiment. They have some plastic boxes to put water in, some graph paper, and a set of measuring tools. They are ready to start.

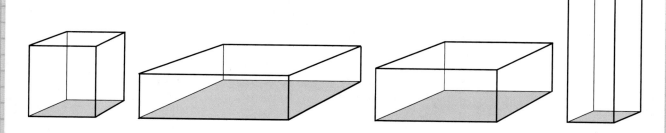

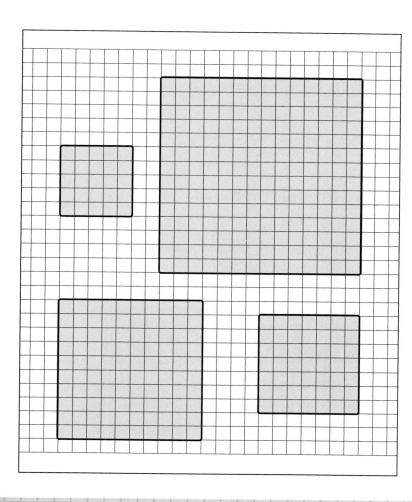

Julie has an idea for measuring the surface area of each box. She traces around each box on graph paper. She uses a meter tape to measure the distance between the lines on the graph paper. The lines are 1 centimeter (cm) apart.

The two students number the boxes. The box with the smallest surface area is number 1. The box with the biggest surface area is number 4. Then they measure 50 milliliters (mL) of water into each box. They place the four boxes on the counter by a window.

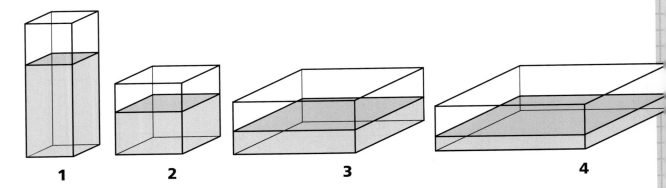

One week later, Julie and Art measure the amount of water in each box. Box 1 has 46 mL, box 2 has 42 mL, box 3 has 34 mL, and box 4 has 18 mL.

Art thinks about the results. It seems that the surface area of the water in the boxes has an effect on the evaporation. But he isn't sure. Julie suggests organizing the results of the experiment. The students decide to do the following.

- Make a T-table to display the data.
- Make a graph of the data.
- Describe what they learned from the experiment.

Can you help Julie and Art? Use the information they gathered to write a report about the effect of surface area on evaporation. Be sure to include the three kinds of information listed above.

Thinking about the Experiment

What additional information would be useful to better understand how surface area affects evaporation?

Condensation

What happens to all that water in the air? As long as the air stays warm, the water stays in the air as water vapor. Warmth (heat) is **energy**. As long as the water vapor has a lot of energy in the form of heat, it continues to exist as a gas.

But if the air cools, things change. As the air cools, the water vapor **condenses**. It changes from a gas into a liquid. When invisible water vapor in the air condenses, the water becomes visible again. Clouds are made of the tiniest droplets of liquid water that have condensed from the air.

Where else have you seen condensation besides up in the sky in the form of clouds? Sometimes water vapor condenses close to the ground. This is called fog. Being in fog is really being in a cloud that is at ground level.

If you go out early in the morning following a warm day, you might see condensation called **dew**. In the pictures below, dew formed on a spider web and along the edges of the leaves on a plant.

Fog close to the ground

Dew on a spider web

Dew on plant leaves

Water vapor condenses indoors, too. On a cold morning you might see condensation on your kitchen window. Or if you go outside into the cold wearing glasses, they could get fogged with condensation when you go back inside.

What happens to the bathroom mirror after you take a shower? The air in the bathroom is warm and filled with water vapor. When the air makes contact with the cool mirror, the water vapor condenses on the smooth surface. That's why the mirror is foggy and wet.

Condensation on a window **Condensation on a mirror**

The Water Cycle

Water particles in the water you drink today may have once flowed down the Ohio River in the Midwest. Those same particles may have washed one of Abraham Lincoln's shirts. They might even have been in a puddle lapped up by a thirsty bison!

Water is in constant motion on Earth. You can see water in motion in rushing streams and falling raindrops and snowflakes. But water is in motion in other places, too. Water is flowing slowly through the soil. Water is drifting across the sky in clouds. Water is rising through the roots and stems of plants. Water is in motion all over the world.

Think about the Ohio River for a moment. It flows all year long, year after year. Where does the water come from to keep the river flowing?

Ohio River

The water flowing in the river is renewed all the time. Rain and snow fall in the Ohio River Valley and the hills around it. The rain soaks into the soil and runs into the river. The snow melts in the spring and supplies enough water to keep the river flowing all summer. Rain and snow keep the Ohio River flowing.

The rain and snow in the Ohio River Valley are just a tiny part of a global system of water recycling. The global water-recycling system is called the **water cycle**.

The big idea of the water cycle is this. Water evaporates from Earth's surface and goes into the air as water vapor. The water vapor condenses to form clouds. The clouds move to a new location. The water then falls to Earth's surface in the new location. The new location gets a fresh supply of water.

A simple water-cycle diagram

Water Evaporates from Earth's Surface

The Sun drives the water cycle. Energy from the Sun heats Earth's surface and changes liquid water into water vapor. The ocean is where most of the evaporation takes place. But water evaporates from lakes, rivers, soil, wet city streets, plants, animals, and wherever there is water. Water evaporates from all parts of Earth's surface, both water and land.

Water evaporates from all of Earth's surfaces.

Water vapor enters the air and makes it moist. The moist air moves up. As moist air rises, it cools. When water vapor in the air cools, it condenses. Water in the air changes from gas to liquid. Tiny droplets of liquid water form. The condensed water is visible. We see condensed water as clouds, fog, and dew.

Water vapor condenses in the air to form clouds.

Water Falls Back to Earth's Surface

Wind blows clouds around. Clouds end up over mountains, forests, cities, deserts, and the ocean. When clouds are loaded with condensed water, the water falls back to Earth's surface as rain. If the temperature is really cold, the water will freeze and fall to Earth's surface as snow, sleet, or hail.

Water falls back to Earth's surface as rain, snow, sleet, or hail.

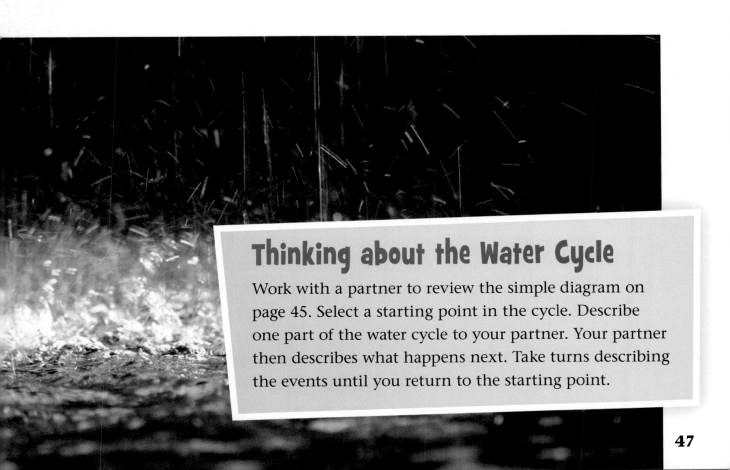

Thinking about the Water Cycle

Work with a partner to review the simple diagram on page 45. Select a starting point in the cycle. Describe one part of the water cycle to your partner. Your partner then describes what happens next. Take turns describing the events until you return to the starting point.

Climate Regions

What's the weather like today? What was it like last year on this date? Probably just about the same. We can predict what the weather will be like next year because weather tends to follow a pattern over a long time. The big patterns of weather describe a region's **climate**.

Climate describes the typical weather conditions in a region. The climate in Hawaii is quite different from the climate in Minnesota. The Hawaiian climate is warm, sunny, and pleasant all year long. The Minnesota climate is freezing cold in the winter, and hot and humid during the summer.

Meteorologists have created models to describe climate in different regions of Earth. The general rule is that a place close to the equator has a warm climate, and a place farther from the equator has a cold climate. With this simple rule, regions fall into three broad groups. The **polar zone** at the North and South Poles has a very cold climate with long winters. The **tropical zone** near the equator has a hot climate and no winter. Everything in between is in the **temperate zone**. With this simple model, most of North America is one climate zone.

This model with just three zones is a little too simple. It doesn't show all of the variations in climate. The most important factors to define a climate zone are the average temperature and the amount of precipitation through the year.

Let's look at a more complex model that has twelve climate zones. We will look at those zones in North America.

Climate Zones

TROPICAL
1. Tropical wet
2. Tropical wet and dry

DESERT
3. Hot arid
4. Interior continental semiarid

TEMPERATE
5. Dry subtropical
6. Humid subtropical
7. Temperate marine

SUBARCTIC
8. Humid midlatitude
9. Subarctic

POLAR
10. Arctic
11. High-altitude
12. Ice cap

Humid Midlatitude Zone

In the Midwest and Northeast, you can be fairly sure that it will be cold and snowy in January and February. It will be warm and humid during the summer. This weather pattern occurs in Minnesota, Illinois, Connecticut, and Maine. The humid midlatitude zone includes the midwestern United States, New England, and the southern part of Canada. This climate zone supports large forests with many different kinds of trees, both evergreen and deciduous.

Humid Subtropical Zone

The southeastern United States rarely have snow in the winter. Spring and summer are rainy, hot, and humid. The humid subtropical zone supports large hardwood forests, palm trees, and many kinds of mosses and vines.

◼ Hot Arid Zone

The hot arid zone in the western United States and parts of Mexico has warm, dry winters. The summers are very hot and dry. Little rain falls during most of the year. During the summer thunderstorms can bring heavy rains that sometimes cause flash floods. The hot arid zone supports many kinds of plants that are adapted for dry conditions, including cactus, mesquite, and yucca.

◼ Interior Continental Semiarid Zone

The semiarid zone has warm spring and summer weather with thunderstorms. The winters are cold. The plants living here include sagebrush and many kinds of grass. This zone stretches from western Canada into Mexico.

High-Altitude Zone

High in the mountains, there are forests of evergreen trees. This high-altitude zone has the right conditions for skiing and other snow sports.

Western mountains, such as the Rockies, Sierra Nevada, and Cascades, are in this zone.

Dry Subtropical Zone

Winter weather in the dry subtropical zone is usually warm and rainy. The summer weather is hot and dry. The dry subtropical zone has oak woodlands and chaparral. It also has a very diverse community of shrubs, grasses, and mixed forests. This zone is excellent for farming, fruit orchards, vegetable gardens, and raising livestock. Parts of central California have this climate.

Temperate Marine Zone

The temperate marine zone of the Pacific Northwest is cool and wet throughout the year. The Pacific Ocean helps create the cool, moist weather. Winters are cool and rainy. Summers are also cool and often foggy. The evergreen forests here grow large redwood, fir, pine, and spruce trees. The moist forests are home to ferns, mosses, lichens, and fungi.

Subarctic and Arctic Zones

Two climate zones occur in most of Alaska and Canada. They are the subarctic zone and the arctic zone. The climate is extremely cold most of the year. The summer is short and cool. Plants are small and close to the ground. This is a result of the very short growing season and harsh winters. Many plants live in wetlands or bogs.

◻ Tropical Wet and ◻ Tropical Wet and Dry Zones

Hawaii has a tropical wet and dry climate. It is warm and sunny all year long, with plenty of tropical rain in many parts of the islands. Other parts of the islands are in rain shadows where there is not much precipitation. The amount of rainfall allows different kinds of vegetation to grow in wet and dry areas.

Climates vary widely across North America. Many states and provinces have only one kind of climate throughout, such as Michigan, Massachusetts, Alabama, and Nova Scotia. Others, such as California and Quebec, have two or more climate zones. So when you are asked what the weather will be like in California, you have to know what part of the state, and what time of the year.

Thinking about Climate Regions

1. Find where you live on the North American climate map. How would you describe the climate in your region?

2. How does the climate in your region change from season to season?

Wetlands for Flood Control

Have you ever seen a **wetland**? Maybe you have without knowing it. Wetlands are also known as bogs, swamps, and marshes. A wetland is an area of land that is partly covered with water. A wetland might be beside a river, a lake, or the ocean. Sometimes a wetland can be dry enough to walk around in. But during times of heavy rain, it can be completely under water.

Wetlands are home to many living things. Many plants and animals that live in wetlands have adaptations that allow them to live in water part of the time and on land part of the time. Wetland plants must be able to live in water in wet times to survive during dry times. Animals like ducks, egrets, and frogs can survive during wet and dry times.

During times of heavy rain, rivers can overflow their banks. The water flows onto large areas of flat, low land next to a river. These lowlands are the river's **floodplain**. Flowing onto its floodplain is a river's way of spreading its extra water over a much larger area. By using its floodplain, a river can hold a large amount of water. That water slowly returns to the river later as the flow shrinks.

You may have never seen a floodplain by a river because people have changed the landscape. Many floodplains have been changed into farmland. Many more have been used as land for building homes and stores. Many floodplains that were once wetlands have been drained of water.

Some floodplains have been changed to farmland.

Often, the water flow in a river is low during the summer, fall, and winter. But in the spring, snow near the river's source begins to melt. The meltwater fills the river. If there is a lot of rain during this time, the rainwater adds to the already heavy flow in the river. This can cause the river to overflow its bank in a **flood**. The extra water can flow into farmlands and cities. The result can be damage to crops, houses, stores, and roads. Floods can also cause the death of large numbers of animals and people. Floods can be very costly natural disasters.

How can people protect themselves from the dangers of floods? For years, people have tried to prevent floods by digging deeper river channels. They have built dams to hold floodwater back. And they have built artificial riverbanks, or levees, to keep high river water flowing in the banks. But these expensive solutions often fail and cause even bigger problems.

Nature has provided some effective defenses against floods. First of all, we must understand that periodic flooding is a natural and healthy part of river ecology. And natural floodplains and wetlands are part of river systems and can reduce water flow. Here's how.

Flood water can cause a lot of damage.

Wetlands can soak up a lot of water. When water floods out of a river channel into a wetland, the wetland soil and plant roots act like a sponge. The wetland reduces the amount of water flowing down the river channel. The wetland holds the water for a while. Then the water slowly seeps back into the river channel. The total flow of water going down the river is not reduced by very much. But the wetlands and floodplains slow the flow rate of the water in the channel. And this reduces the effects of the water as it flows downriver. Wetlands and floodplains help to reduce erosion and lower the level of floods.

The wetland plants slow the rate of flow in another way. Plants such as rushes and grasses provide barriers. The water must flow around and over the plants. This reduces the speed of water flowing down the main river channel.

Wetland plants, such as rushes and grasses, act as a water barrier.

A storm surge can produce huge waves that flood coastal areas.

Rivers aren't the only sources of flood hazard. Ocean water is subject to tides. Every day the tide comes in and then goes out. When the highest tides flow in, low coastal areas might flood a little bit with sea water. But when large tropical storms come near shore, the strong winds can blow an extra load of sea water up onto the land. This kind of flood wave is called a **storm surge**. A storm surge wave can be several meters high. The surge can push water into towns along coastal areas. It can destroy buildings and leave several meters of salty water on the land.

Again, one of the best defenses against this kind of flood is a wetland. Wetlands exist along many parts of the coast, particularly at the mouths of rivers and creeks. Rushes and reeds, mangrove and cypress trees, and salt-tolerant grasses grow in the wetlands. These wetland plants form a natural barrier against the rushing surge water. Slowing the surge waters reduces the height of the surge. It reduces the force of its impact. And the wetlands limit how far the water flows inland.

Wetlands also trap silt and sand, creating offshore land. In some places, this process creates **barrier islands**. Barrier islands are strips of narrow land a short distance from shore. They provide a large buffer zone between the open ocean and coastal cities. The only problem is that people like to build their homes on these beautiful islands. These homes have no buffer against the full force of surges when tropical storms strike.

Some climatologists think that tropical storms are getting more powerful. These storms may bring heavier than usual rainfall and produce more frequent floods. It would be wise for people living in flood zones to make plans to protect their wetlands. City planners, farmers, geologists, **engineers**, biologists, teachers, and students can work together to preserve wetlands and floodplains. The result would reduce the damaging effects of these floods.

Thinking about Wetlands for Flood Control

1. Have there been floods in your community? What caused them? What was the effect of the floods?

2. Is there a floodplain in your community? How is it used?

3. Is there a wetland in your community? How does that wetland help reduce the effects of floods?

Conserving Water during Droughts

Have you ever lived in a part of the country or the world that experienced a drought? A drought is a shortage of water caused by below-average rainfall. It can be scary, because water is needed for life. During a drought, every bit of water needs to be conserved.

In many places, drinking water is stored in large **reservoirs** until it is used. A research team at the Massachusetts Institute of Technology (MIT), led by Moshe Alamaro, tried to find a way to conserve water by reducing evaporation from reservoirs.

Dr. Moshe Alamaro

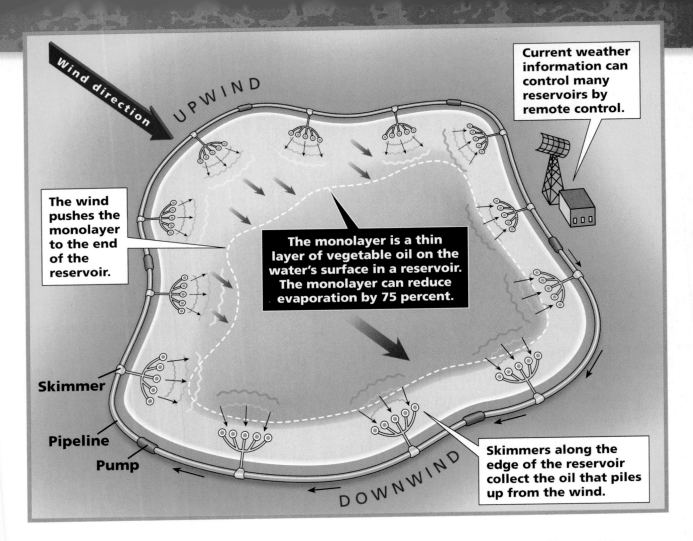

Wind direction

UPWIND

Current weather information can control many reservoirs by remote control.

The wind pushes the monolayer to the end of the reservoir.

The monolayer is a thin layer of vegetable oil on the water's surface in a reservoir. The monolayer can reduce evaporation by 75 percent.

Skimmer

Pipeline

Pump

DOWNWIND

Skimmers along the edge of the reservoir collect the oil that piles up from the wind.

One way to reduce evaporation of water is to cover it up. But putting a lid on a large reservoir would be very expensive.

Alamaro came up with a different design. His plan involves placing a very thin layer of vegetable oil on the surface of the water. This layer is called a monolayer because it is one oil particle thick. The monolayer is made of olive, palm, or coconut oil, so it is safe for animals. If the monolayer stays in place, it can reduce evaporation rates by about 75 percent. That's a lot of water saved!

The idea of reducing evaporation with oil is not new, but the idea hasn't worked because of wind. The wind pushes the monolayer to one end of the reservoir. Alamaro is working to solve this problem. His idea is to place skimmers around the perimeter of the reservoir. These skimmers would collect the oil that piles up, and then pump it upwind through pipes. He's working in California, Texas, and Massachusetts to try this on a large scale. If it works, droughts might be just a little less scary.

Could you set up an experiment to see if oil reduces evaporation? Give it a try!

WATER: A Vital Resource
by Keira, David, Tamiko, and Jorge

Our team's assignment was to learn about our water supply. Earth seems to have plenty of water. But 97 percent of that water is salt water. Another 2 percent of the world's water is frozen. That leaves just 1 percent as fresh water. The good news is that 1 percent should be enough for everyone. The bad news is that it's not spread equally around the world. Some places have a lot of fresh water, but others do not.

Tamiko brought some information to our first meeting. She said we use 35 times more water today than people did 300 years ago. The human population has grown, and so have the ways we use water.

Jorge looked surprised and asked, "Will we ever run out of water?" Keira was sure the answer was no. She reminded us about the water cycle. She said, "The amount of water on Earth doesn't get used up. It gets recycled."

But David wondered if the amount of water we need will grow larger than the amount we have. That made all of us realize how important it is to take care of the water we have. Tamiko summed it up this way. "Water is one of the most valuable natural resources on Earth! We have to take care of it. If we make our water too dirty to use, or if we use our water faster than it is replaced, we could be in a lot of trouble."

At the end of the meeting, each of us chose something to investigate about our water supply. We agreed we had a lot to learn.

Water is a renewable resource, but it is not unlimited.

Water Coming into Our Homes
by Keira

My community takes water from Lake Charles. In other places, water comes from rivers or underground **aquifers**. Aquifers contain water that has soaked into the ground and is stored in layers of rock. Water is usually treated before it reaches our faucets. Water-treatment plants filter and treat the water, making it clear and safe for people to use.

First, water is screened to remove fish, leaves, and large objects such as logs or trash. Next, a machine called a flash mixer stirs the water with chemicals. Four chemicals are commonly mixed with water. They are lime, carbon, chlorine, and alum. Lime softens the water. Carbon **absorbs** materials that smell bad. Chlorine kills bacteria. Alum makes particles of clay clump together.

The mixed water goes to a settling tank. Clay clumps, silt, and other particles drift to the bottom. From there, the water passes through sand, gravel, and charcoal filters. A chemist at the water-treatment plant tests the water every day. This is to make sure that all harmful bacteria are killed. Purified water is pumped into tanks and towers. It reaches our homes through underground pipes.

Water Leaving Our Homes
by David

Water leaves our homes. It runs down sink and bathtub drains and out of washing machines. It is flushed down toilets. Waste water must be treated before it returns to the environment. In some communities, waste water goes to sewage-treatment plants. In other places, waste water enters local septic systems.

Septic tanks are usually made of concrete or metal. They are buried outside houses. Waste water separates inside a septic tank. Heavy materials sink to the bottom and form sludge. Lighter materials like fats and grease rise and form scum. Bacteria break down solids in the tank. The liquid in the middle flows through pipes into gravel-filled trenches. The liquid in the trenches is purified as it seeps through the gravel and soil.

Sewage-treatment plants screen waste water to remove solids. Bacteria break down other materials. Chemicals are used to rid the water of impurities. Treated water then discharges into streams, lakes, or the ocean.

City Runoff
by Tamiko

Rain does not always evaporate or soak into the ground. Sometimes it becomes **runoff**. Runoff flows over land and streets, and then into storm drains. Storm drains often empty right into bays, lakes, and streams.

Some people don't know that storm drains connect to local water systems. They sometimes pour pet waste, oil, paint, and other hazardous materials into storm drains. The untreated water harms the water supply.

In some cities, science clubs or environmental groups paint pictures of fish on the sidewalks near storm drains. The fish remind people that whatever goes into the storm drain will enter the water supply.

Water Conservation
by Jorge

Conserving water is an important part of protecting it. Because of conservation, water use in the United States has dropped since 1980. Here are a few things we can do to save water.

- Turn off the tap when you brush your teeth. Don't run water while washing dishes. Shut the shower off while you soap up.

- Take shorter showers and use a low-flow showerhead.

- Install low-flow aerators on all your faucets. An aerator mixes air with the water. You use less water when air is mixed in. The flow will still seem strong.

- Fix leaks in pipes, faucets, and toilets. Dripping faucets can waste about 7,500 liters (L) of water each year. Leaky toilets can waste as much as 750 L each day.

- Use less water to flush your toilet. Install a low-flow toilet, or put a water-filled plastic container in the tank if you have an older toilet.

- Use a broom, not a hose, to clean driveways and sidewalks.

- Water lawns and other outdoor plants in the morning. (Water evaporates faster in the middle of the day.) Don't water on a windy day.

- Put mulch around plants to reduce evaporation.

Thinking about Water

1. What is the source of your local water?

2. How is water purified in your community?

3. What are the issues about water in your community?

Natural Resources

Some people call it "dirt." Others call it "earth" or "the ground." What they are talking about is soil. Soil is the layer on top of the land. Soil is what you dig up with a shovel. You can stir soil with water to make mud or turn it over with a plow.

The soil in your schoolyard is different from the soil in a field. The soil in a field is different from the soil in a desert. In fact, soils are different just about every place you look. But in some ways, soils all over the world are the same.

All soils have two basic ingredients: rock and **humus**. The rock part of the soil comes in a variety of sizes, including gravel, sand, silt, and clay. Particles of gravel are rocks the size of rice and peas. Sand particles are smaller rocks. Silt particles are so small it's impossible to see just one. Clay particles are smallest of all.

Humus is black **decomposing organic matter**. It comes from the dead and discarded parts of plants and animals.

Soil in a field prepared for planting

Soil in the Mojave Desert

But soils have different **properties**. They differ in **texture** and color. Soils also differ in their ability to **retain** water and to support plant growth. The texture of soil depends on the amount and size of the rock particles. Soils with a lot of sand and gravel feel gritty. They fall apart easily when you make a mud ball. Soils with a lot of silt and clay feel smooth and slippery. They make excellent mud balls.

Mud balls made out of sandy soil and clay soil

Soil color depends on the color of the sand, silt, and clay particles, and the amount of humus. Texture and humus determine the amount of water that a soil can retain. If the soil has a lot of sand and gravel particles, water will flow through the spaces between the particles. Very little water will stay in the soil. Soil with smaller particles and a lot of humus will retain more water. The water gets trapped in the smaller spaces between particles. And water is absorbed by the organic humus particles.

Soil is a renewable resource.

Soil as a Natural Resource

Materials that people get from the natural environment are natural resources. Most of the food that people eat comes from plants or from animals that eat plants. Plants grow in soil. The soil is a natural resource that people depend on for survival.

Soil is a **renewable resource**. That means that natural processes make new soil, but it happens very slowly. Soil must be used wisely. If the soil resource is overused, it will lose its ability to support the growth of plants. Farmers need to renew the soil nutrients to make sure plenty of food crops will be available for people.

Other Natural Resources

People rely on many other natural resources. There are renewable resources and **nonrenewable resources**.

Renewable resources are replaced as we use them. We have investigated one important natural resource, water. We know that water is renewed all the time by the water cycle. Plants and animals are also renewable resources. New plants and animals are growing all the time. We use them for food and shelter. Wood is another example of a renewable plant resource.

When nonrenewable resources are used up, they are gone. People use a lot of nonrenewable natural resources as **energy sources**. Coal, petroleum, and natural gas are energy sources. These **fossil fuels** are the remains of plants and animals that lived millions of years ago. When Earth's fossil fuels are used up, they will be gone forever. No new fossil fuels are "growing" at this time. The length of time that we have fossil fuels can be extended by conservation. People can conserve fossil fuels by using more energy-saving products, like high-mileage cars and better insulated homes.

Lumber is a renewable resource.

Petroleum and coal are nonrenewable resources.

71

Some natural resources are **perpetual renewable resources**. Perpetual means they are available all the time whether we use them or not. Examples of perpetual renewable resources are solar energy, wind power, geothermal energy, and tides. The most important energy sources for the future will be based on energy from the Sun.

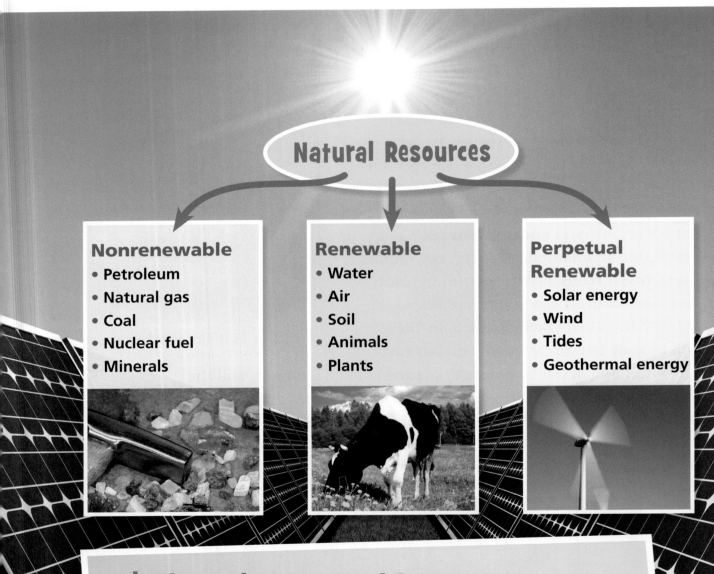

Natural Resources

Nonrenewable
- Petroleum
- Natural gas
- Coal
- Nuclear fuel
- Minerals

Renewable
- Water
- Air
- Soil
- Animals
- Plants

Perpetual Renewable
- Solar energy
- Wind
- Tides
- Geothermal energy

Thinking about Natural Resources

1. Select one nonrenewable resource. Explain why it is considered nonrenewable. What can people do to conserve this resource?

2. Select one renewable resource. Explain why it is considered renewable. What can people do to conserve this resource?

3. Describe what is meant by perpetual renewable resources.

4. Explain why it is important to conserve all natural resources.

Ellen Swallow Richards: An Early Ecologist

An American in 1900 could expect to live only to age 47. Today life expectancy is much longer. We owe that in part to Ellen Swallow Richards. She lived in a time when people understood little about germs and pollution. Yet Richards knew there was a connection between health and the environment. In the early 1900s, she wrote to the president of the Massachusetts Institute of Technology (MIT), "One of the most serious problems of civilization is clean water and clean air, not only for ourselves but for the world."

Ellen Swallow was born on December 3, 1842. She lived in Dunstable, Massachusetts. Growing up, Ellen did chores on her family's farm and helped in their store. She also took care of her mother, who was often sick. Ellen's first teachers were her parents. They saw that Ellen loved to learn. Before long, the family moved to Westford, Massachusetts, where Ellen entered school.

Ellen became a teacher after graduation. When her mother became ill again, Ellen returned home to help. But she was unhappy working in the family store. She wanted to learn more, and she wanted to go to college.

Richards in her study

Few colleges accepted women at that time. Many people believed studying hard would make women ill! But Ellen would not forget her dream. She worked at many jobs and saved all the money she could. Finally she had enough money to enter Vassar College. Vassar was an experimental school. It aimed to give women the same chance that men had to get an education.

Ellen was called a "special student" at Vassar because she was 26 years old. The other women were 14 to 19 years old. Ellen was too happy to care. Her favorite subjects were astronomy and chemistry. In 1870, she was part of Vassar's first graduating class.

Ellen planned to teach in Argentina, but war broke out. Instead she entered graduate school at MIT. She was not charged tuition. Ellen believed this was because she was poor. In fact, MIT was afraid to admit women. By not charging Ellen, the school could claim she was not really a student.

Ellen worked at MIT after her graduation in 1873. The professors respected her. One laboratory head said, "When we are in doubt about anything, we always go to Miss Swallow." Ellen married chemistry professor Robert H. Richards in 1875. They helped each other with their work.

Richards collecting water samples

In 1884, Ellen Swallow Richards became an instructor of "sanitary chemistry." For 2 years, she and Professor Thomas M. Drown studied the state's water supply. They suspected that something in the water was making people sick. Richards worked to find a way to test the **water quality**. Water was collected from every river and lake in Massachusetts once a month. Richards analyzed most of the 40,000 samples herself. When the survey was done, Massachusetts had the first standards for water purity. Professor Drown wrote that this was "mainly due to Mrs. Richards's great zeal and vigilance." From then on, Richards taught others how to analyze air, water, and sewage.

Ellen Swallow Richards started the Women's Laboratory at MIT in 1876. She wanted other women to study science. When the Women's Laboratory closed in 1883, Richards was thrilled. Through her efforts, women were no longer "special" at MIT. They were regular students, equal to men.

Ellen Richards with female students in 1888

Another of Richards's interests was nutrition. She opened the New England Kitchen, where immigrants were taught how to cook nutritious, inexpensive food. She cared deeply about public health. She urged women to eat right and exercise.

Ellen Swallow Richards died at the age of 68 on March 30, 1911. Many people consider her to be the founder of ecology. She said, "The quality of life depends on the ability of society to teach its members how to live in harmony with their environment." It was her belief that science should make people healthier. She worked hard to make that happen.

The MIT chemistry department in 1900

Making Drinking Water Safe

What happens if you are outside playing at recess and you get thirsty? You walk over to a drinking fountain for a drink. And if your dad needs to boil some water to make dinner, he goes to the sink and turns on the faucet.

In the United States, we don't usually think about how easy it is to get water that's safe to drink. We just turn on the faucet, and out comes clean water. Our water is treated and tested for safety before it gets to us.

What about other countries? Many people don't have running water in their homes and schools. They go to lakes and rivers to get water. Sometimes the water contains bacteria. Most bacteria don't hurt people, but some can make people sick. Boiling kills bad bacteria, but not everyone can boil their water whenever they need to. Many engineers around the world are seeking solutions to problems related to clean water.

Solar Disinfection System

In 1991, a group of scientists and engineers in Switzerland tried to use everyday tools to make drinking water safe. They wanted to find a way to get rid of disease-causing bacteria without expensive chemical treatment.

The solution was a solar disinfection system, or SODIS. It relies on a few simple items and sunshine. SODIS works best in countries near the equator. That's where sunshine is strongest.

Here's how SODIS works.

1. Get a clean, clear plastic bottle with a cap. (Glass can be used, but clear plastic is best.)
2. Fill the bottle with water and put the cap on.
3. Lay the bottle flat on a piece of corrugated tin or on a roof.
4. Let the bottle lie in the sunshine for 6 hours to 2 days, depending on how cloudy it is. Then the water is ready to drink.

Light from the Sun is called **solar radiation**. The ultraviolet part of solar radiation kills bad bacteria and makes it safe to drink. Heat and ultraviolet radiation from the Sun work together to disinfect the water.

What's so great about SODIS? It costs nothing at all, and it recycles plastic bottles. Sometimes the simplest solutions are the best.

1. Clean the bottles.
2. Fill the bottles with water.
3. Put the bottles in sunshine.
4. Wait 6 hours to 2 days.
5. Drink the safe water.

These pots are ceramic water filters engineered to remove bad bacteria.

Ceramic Water Filters

Vinka Oyanedel-Craver is an environmental engineer at the University of Rhode Island. She is working to help people in developing countries create safe drinking water right in their own homes. For several years, she has tested and improved water filters that look like flowerpots without a hole on the bottom. One ceramic pot can slowly filter up to 7 liters (L) of dirty water at a time to remove sediment and dangerous bacteria.

When you pour water into the pots, it slowly filters through the porous walls of the pot. This filtering is similar to what water would do if you poured it onto a sponge. The pots are made from clay and very fine sawdust. When the pots fired in a kiln, the sawdust burns. This leaves microscopic holes in the pot. If you looked at the clay with a very powerful microscope, you would see these holes. The tiny holes allow pure water to pass through, and trap the things that you don't want in your drinking water.

The pots alone clean the water pretty well, but they don't get out all of the dangerous bacteria. Silver is a disinfectant that can kill dangerous bacteria. Potters thought if they painted the pots with a thin layer of silver, it would make the water even safer to drink. Before the engineers got involved, the potters were painting their pots with silver nitrates. When Oyanedel-Craver tested these pots, she discovered that 40 percent of the silver nitrates came off the first time the filter was used. This was wasteful and potentially dangerous. After much testing, Oyanedel-Craver and others discovered that painting the pots with ultrafine silver particles was less wasteful. This new paint worked better to remove most of the bacteria.

Painting the pots with silver kills more bacteria.

Vinka Oyanedel-Craver is an environmental engineer who grew up in Chile. She discovered in high school that she wanted to be an engineer. One of her teachers told her that he thought she would be a good engineer. Today, she makes a difference in the communities that use these filters. Because of this work, many people in these communities have jobs making the pots. And everyone who depends on these filters can trust that the water from the filter is clean and safe to drink.

Making ceramic water filter pots

Removing Arsenic

In other parts of the world, there are different problems with drinking water. Susan Amrose (1977–), is an environmental engineer at the University of California, Berkeley. She has worked with communities in Bangladesh where most of the **groundwater** coming from wells is poisoned with naturally occurring arsenic. Arsenic is too small to filter out of water. Most people know the water is dangerous, but they don't have anything else to drink.

Amrose works with a team of engineers to design an inexpensive way to remove the arsenic. One method is electro-chemical arsenic remediation (ECAR). First, iron pieces are added to the water. Then, electricity is run through the iron. This speeds up the rusting of the iron. The rust particles dissolve in the water. The tiny arsenic particles attach to the rusting iron and make much bigger particles. Finally, the bigger particles can be filtered out to make the water safe to drink.

The iron is toxic after the arsenic attaches to it. After this waste product is filtered out of the water, it is mixed into concrete and used for roads. When the iron and arsenic are in the concrete, they are very safe and can't get back into the environment.

Susan Amrose

The ECAR equipment used to remove arsenic from water

Amrose didn't become an engineer right away. In fact, she started working on her PhD in astrophysics. After taking many physics classes, she took her first engineering class. It changed her career, and she ended up looking at water treatment.

Amrose has a strong connection to this project. "I've visited families in Bangladesh who only have arsenic-contaminated water sources for drinking. A father once asked me to please hurry up my research if I could, because he wanted his young daughter to grow up and thrive. He said it was too late for him (because he had been drinking arsenic-contaminated water his whole life and had many health problems), but he wanted his children to survive and have a happy life."

Engineers

The engineers featured in this article are doing amazing things to make the world a better place to live. And they're all engineering with water. Maybe you will decide to be an engineer. What problems will you help solve?

Using the Energy of Water

The water cycle moves water all over Earth. Energy from the Sun evaporates water and lifts it high in the air. The water condenses into clouds. Wind moves clouds all over Earth. Eventually the water falls from the clouds as rain, snow, sleet, or hail.

A lot of water falls high in the mountains. Water is matter. We know what happens to matter on a slope. The force of gravity moves it downhill. When water runs into something, it applies a force. Moving water has the force to push things around.

During very heavy rainstorms, rivers and streams can flood and overflow their banks. The force of the flood water can wash away rocks and soil, destroy roads, and carry away cars and houses. The faster water flows, the more force it has, and the more damage it can do.

Hurricanes are strong storms that produce extremely high winds. When hurricanes come on land, they can cause a storm surge. A storm surge is a huge wall of water that washes onshore. On August 29, 2005, Hurricane Katrina hit New Orleans with a huge surge. The force of the surge plus the flow of the Mississippi River broke through the levees protecting the city. The resulting flood caused huge devastation. More than 1,800 people died, and the estimated cost of the damage was more than $100 billion.

A flood following heavy rain washes out a road.

The flood following Hurricane Katrina did massive damage.

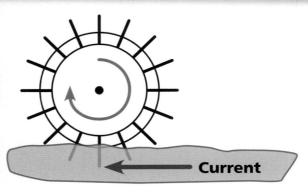

Current

Using Water to Do Work

Were you able to put water to work in class? **Waterwheels** have been used for thousands of years. The early Greeks and Romans used them to grind grain. Early American towns used waterwheels to power gristmills and sawmills.

Two different forces can push or pull on a waterwheel. Moving water makes it turn. The force from the stream of fast-moving water hitting the **blades** pushes the shaft around. In this old-fashioned waterwheel, the current in the stream pushes on the blades at the bottom of the wheel to turn it.

Another way to drive a waterwheel is to fill "buckets" attached to the outside of a wheel. Water pours onto the top of the wheel. The buckets catch the water. The weight of the water pulling down turns the wheel. The water in the buckets spills out as the wheel turns.

Water

Current

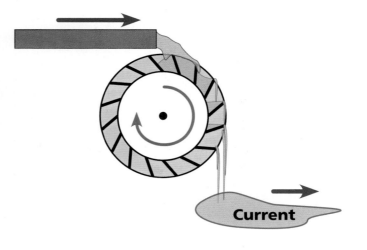

Current

Huge generators are built into
the base of Hoover Dam in
Arizona and Nevada.

A modern kind of waterwheel is the
water turbine. Water turbines are built
into the bottoms of dams. Water from the
reservoir behind the dam turns the turbine
to generate **electricity**.

High-pressure water flows into a chamber
above the turbine. Water then flows through
the turbine, pushing on the turbine blades
as it goes. The blades turn a shaft, which is
connected to the generator.

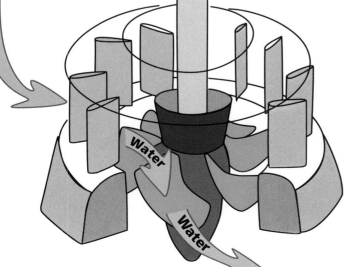

Water pushes the
blades as it flows
through the turbine.

Water Engineers

Hydroelectric dams have been built on many rivers around the world. Water flows through them and spins turbines to generate electricity. The dams change the landscape upstream, and this affects all the living things that live in and around the river.

Today, many people are thinking about "green" sources of electricity. When people say things are green, they aren't talking about the color. They're thinking of things that do not harm the environment. One engineering challenge is to design a device that will harness energy in slow-moving water in rivers and the ocean. One **criterion**, or need, of the design is that it must not hurt living things in the water.

Michael Bernitsas (1952–) is an engineer working on green electricity generation at the University of Michigan. He has spent several years working on an invention that uses the power of slow-moving water to generate electricity.

Dr. Michael Bernitsas

Bernitsas' device is a hydro energy converter. It rests on the bottom of a river or ocean bay. Gravity holds it in place. The large, heavy, open box has several cylinder shafts. The cylinders are positioned a bit like the rod was positioned in your waterwheel. The flow of the water makes the cylinders move up and down instead of round and round. This movement creates electrical power.

What makes the cylinders move up and down? Have you ever pulled a canoe paddle through the water? If you look closely, you might see a whirlpool spin off the end of the paddle. That whirlpool is a vortex.

Laboratory testing for the hydro energy converter

As the river or bay water flows over the cylinders, it creates a vortex. The vortex changes the pressure on one side of the cylinder, causing the cylinder to move away from the vortex. As a result, the cylinders move up. When another vortex comes, they move down. The up-and-down movement of the cylinders is used to generate electricity. This is the process that Bernitsas and other engineers are using to generate electricity.

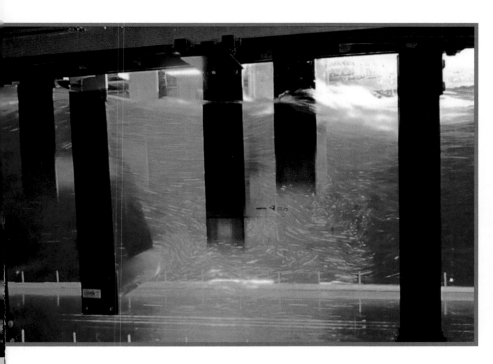

Can you see a vortex in the testing tank? How is this photo like the other laboratory photo on this page? How is it different?

This system generates electricity in slow-moving water as well as in faster-moving water. Most of the currents in rivers and the ocean around major population centers are slow. The machines are affordable and reliable. They are silent and invisible to everyone on land. Fish can safely swim among the cylinders, so there is little impact on the environment. The engineers are testing the device in a river to collect data about its use in the environment.

"Beta 1" was first tested on the St. Clair River near Port Huron, Michigan. It worked! Now machines are being built for testing in other parts of the world. Engineers are looking for ways to make the machine more efficient. They are studying the way that fish swim closely together in schools. Bringing the cylinders on the device closer together might improve efficiency. Making the machines more efficient will make electricity less expensive.

When the cylinders are vertical as shown here, they do not float to the surface of the water, and they produce more electricity.

Many engineers work on issues related to water. Some engineers design devices to harness energy. Others work on new designs for boats to transport things across the ocean. Others design systems for capturing rainwater, or design ways to make water safe for drinking. Other engineers design water-efficient toilets, faucets, and showerheads that use less water.

When you designed a waterwheel in class, you were doing the work of an engineer. There are many types of engineers. Mechanical engineers develop machines, tools, and equipment. Electrical engineers design circuits to power devices and the components to generate and deliver power. Civil engineers work to improve the living and working environment. They work on highways, bridges, and water and sewage systems. Materials engineers develop new materials that can be used to construct new products. Engineering is a creative process. The main job of engineers is to design products, processes, or systems to meet the needs of people.

Science Safety Rules

1. Listen carefully to your teacher's instructions. Follow all directions. Ask questions if you don't know what to do.

2. Tell your teacher if you have any allergies.

3. Never put any materials in your mouth. Do not taste anything unless your teacher tells you to do so.

4. Never smell any unknown material. If your teacher tells you to smell something, wave your hand over the material to bring the smell toward your nose.

5. Do not touch your face, mouth, ears, eyes, or nose while working with chemicals, plants, or animals.

6. Always protect your eyes. Wear safety goggles when necessary. Tell your teacher if you wear contact lenses.

7. Always wash your hands with soap and warm water after handling chemicals, plants, or animals.

8. Never mix any chemicals unless your teacher tells you to do so.

9. Report all spills, accidents, and injuries to your teacher.

10. Treat animals with respect, caution, and consideration.

11. Clean up your work space after each investigation.

12. Act responsibly during all science activities.

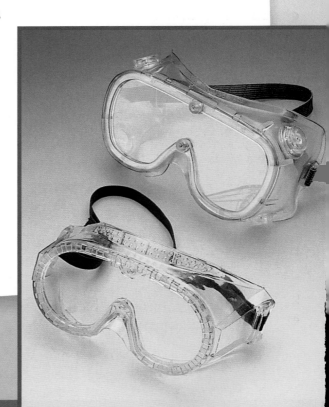

Glossary

absorb when a liquid soaks into a material

air pressure the force exerted on a surface by the mass of the air above it

anemometer a weather instrument that measures wind speed with wind-catching cups

aquifer water that is underground in layers of rock or sediment

barrier island a strip of narrow land a short distance from shore

blade the part of a waterwheel that the water pushes as it moves downward

boiling point (100°C) the temperature at which water changes to gas

Celsius (°C) the basic unit of temperature in the metric system. Water freezes at 0°C and boils at 100°C

climate the average or typical weather conditions in a region of the world

cloud tiny droplets of water, usually high in the air

compass a magnetic needle in a case. Compass needles on Earth point north.

condense when water vapor touches a cool surface and becomes liquid water

conserve to use carefully and protect

contract to get smaller; to take up less space

criterion (plural **criteria**) a need or requirement

decomposing organic matter humus; dead or discarded parts of plants and animals

density the amount of mass compared to the volume

dew water that condenses on a surface when the temperature drops at night

drought a less-than-normal amount of rain or snow over a period of time

electricity energy that flows through circuits and can produce light, heat, motion, and sound

energy the ability to make things happen. Energy can take a number of forms, such as heat and light.

energy source a place where energy comes from, such as coal, petroleum, and natural gas

engineer a scientist who designs ways to accomplish a goal or solve a problem

evaporate when liquid water in a material dries up and goes into the air

evidence data used to support claims. Evidence is based on observation and scientific data.

expand to get bigger; to take up more space

float to stay on the surface of water as a result of being less dense than water

flood a large amount of water flowing over land that is usually dry

floodplain the flat, low land area next to a river that may flood

force strength or power exerted on an object

forecast to predict future events or conditions, such as weather

fossil fuel the preserved remains of plants and animals that lived long ago and changed into oil, coal, and natural gas

freeze to change from a liquid to a solid state as a result of cooling

freezing point (0°C) the temperature at which water becomes a solid (ice)

fresh water water that is in lakes, rivers, groundwater, soil, and the atmosphere

gas a state of matter with no definite shape or volume; usually invisible

glacier a large mass of ice moving slowly over land

gram (g) the basic unit of mass in the metric system

gravity the natural force that pulls objects toward each other. On Earth, all objects are pulled toward the center of Earth because of gravity.

groundwater water found in the spaces between rock particles (sand, gravel, pebbles), and in cracks in solid rock

heat observable evidence of energy

humidity water vapor in the air

humus (HEW-mus) bits of dead plant and animal parts in the soil

hurricane a severe tropical storm that produces high winds

ice the solid state of water

iceberg a large mass of ice that has broken from a glacier and floats in the ocean

liquid a state of matter with no definite shape but a definite volume

Liter (L) the basic unit of liquid volume in the metric system

mass the amount of material in something

matter anything that has mass and takes up space

measure to compare the size, capacity, or mass of an object to a known object or known system

melt to change from a solid to a liquid state as a result of warming

meteorologist a scientist who studies the weather

mixture two or more substances together

more dense when an object has more mass for its size than another object. When an object sinks in water, it is more dense than water.

natural resource a material such as soil or water that comes from the natural environment

nonrenewable resource a natural resource that cannot be replaced if it is used up

opinion a claim based on belief, not on scientific data or observations

perpetual renewable resource a renewable resource that lasts forever

polar zone a very cold climate with long winters (North and South Poles)

precipitation rain, snow, sleet, or hail that falls to the ground

predict to estimate a future event based on data or experience

property something that you can observe about an object or a material

rain liquid water that is condensed from water vapor in the atmosphere and falls to Earth in drops

recycle to use again

renewable resource a natural resource that can replace or replenish itself naturally over time

reservoir a place where water is collected and stored

retain to hold or continue to hold

runoff rain that does not evaporate or soak into the ground

salt water ocean water

scale something divided into regular spaces to use as a tool for measuring. Rulers and thermometers are both scales.

shaft a rod or bar that rotates

sink to go under water as a result of being more dense than water

soak to be absorbed or move into another material

soil a mixture of humus, sand, silt, clay, gravel, or pebbles

solar radiation light from the Sun

solid a state of matter that has a definite shape and volume

speed the measure of an object's change in position over time

storm surge when water piles up along a coast, rushing toward land faster than it can return to sea

surface area the area of liquid exposed to or touching the air

surface tension the skinlike surface on water (and other liquids) that pulls it together into the smallest possible volume

temperate zone the climate for the majority of Earth, which includes a wide range of temperatures

temperature a measure of how hot or cold the air is

texture the feel or general appearance of an object or a material

thermometer a tool used to measure temperature

tropical zone a hot climate with no winter

volume three-dimensional space

water a liquid earth material made of hydrogen and oxygen

water cycle the repeating sequence of condensation and evaporation of water on Earth, causing clouds and rain and other forms of precipitation

water quality a term used to describe the purity of water

water turbine a modern waterwheel

water vapor the gaseous state of water

waterwheel a wheel turned by the force of moving water

weather the condition of the air around us

weather balloon a balloon that carries weather instruments into the sky

weigh to find the mass of. An object is weighed to find its mass.

wetland an area of land close to a large body of water

wind meter a weather instrument that measures wind speed with a small ball in a tube

wind vane a weather instrument that measures wind direction

Index